BIORHYTHMS

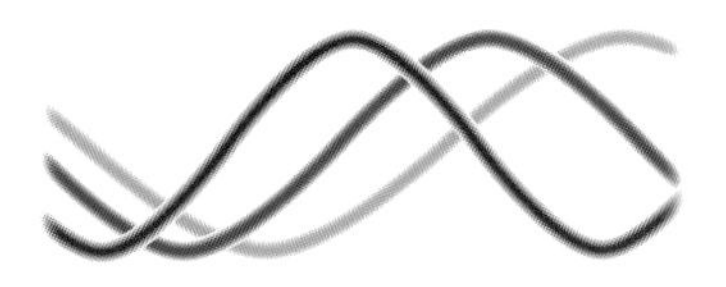

IN A NUTSHELL

BIORHYTHMS

A STEP-BY-STEP GUIDE

PETER WEST

ELEMENT

SHAFTESBURY, DORSET • BOSTON, MASSACHUSETTS • MELBOURNE, VICTORIA

First published in Great Britain in 1999 by
ELEMENT BOOKS LIMITED
Shaftesbury, Dorset SP7 8BP

Published in the USA in 1999 by
ELEMENT BOOKS INC.
160 North Washington Street,
Boston MA 02114

Published in Australia in 1999 by
ELEMENT BOOKS LIMITED
and distributed by
Penguin Australia Ltd.
487 Maroondah Highway,
Ringwood, Victoria 3134

NOTE FROM THE PUBLISHER
Any information given in this book is not intended to be taken as a replacement for medical advice. Any person with a condition requiring medical attention should consult a qualified practitioner or therapist.

Designed and created for Element Books with
The Bridgewater Book Company Limited

ELEMENT BOOKS LIMITED
Managing Editor Miranda Spicer
Senior Commissioning Editor Caro Ness
Editor Finny Fox-Davies
Group Production Director Clare Armstrong
Production Manager Susan Sutterby
Production Controller Claire Legg

THE BRIDGEWATER BOOK COMPANY
Art Director Sarah Howerd
Designer Jane Lanaway
Editorial Director Sophie Collins
Editor Andrew Kirk
Picture research Lynda Marshall

Printed and bound in Singapore by
Tien Wah Press Pte Ltd.

Library of Congress Cataloging in Publication
British Library Cataloging in Publication
data available

ISBN 1 86204 478 3

The publishers wish to thank the following for the use of pictures:
The Freud Museum, The Hulton Getty Picture Collection, Image Bank, The Stock Market, Telegraph Colour Library.

Special thanks go to:
Clare Bayes, Andrew Brown, David Burton, Guy Corber, Maia, Ozora, Hasia, and Aaron Curtis, Carly Evans, Helen Furbear, Simon Holden, Wendy Oxberry, Patricia Sawyer, Emma Scott *for help with the photography.*

Contents

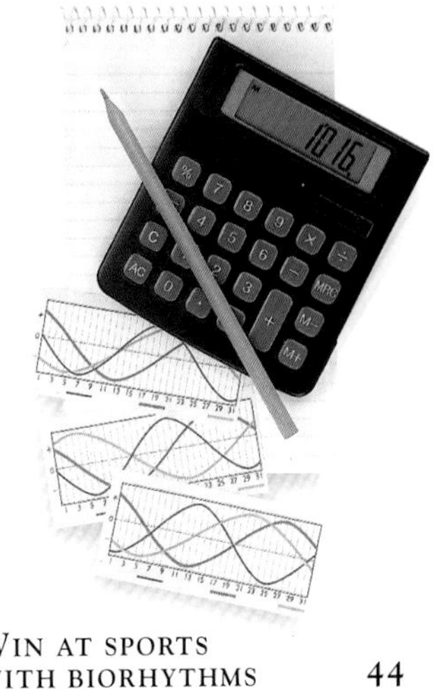

What are biorhythms?

BIORHYTHMS ARE *continuous physiological changes that recur in a series of never-ending measurable cycles within our bodies. Awareness of them can help you to plan your life and live it more positively. It is important to understand that these cycles, irrespective of their phases, have little apparent cause and effect, though biorhythms occasionally influence behavior patterns to a greater or lesser degree depending on the individual.*

THE EFFECTS OF BIORHYTHMS

Although there are three main inner body rhythms to consider, there are also a host of other cycles that influence us constantly from inside and from outside our bodies. Some we recognize and we adapt to them. Others we are unaware of and remain ignorant about all our lives. These rhythmic episodes are with us all the time however, and, while we can do little about their regularity, we can and do adapt to them.

There is little we can do about these biorhythmic forces. Take, for example, the obvious cycle of night and day. We have adapted to its pattern in ways we do not always fully appreciate. Many ignore these cycles by harnessing other energies such as light and power that

BELOW ***Working late means resisting normal rest patterns.***

enable them to work at night, overriding normal sleep patterns. Because the cycles of spring, summer, fall, and winter cannot be ignored, we have adapted to them, too. At sea, tidal influences are disregarded at our peril and we are forced to plan accordingly. Once we recognize the full influence of unseen forces in our lives we can learn when and how to exploit them for our own advantage.

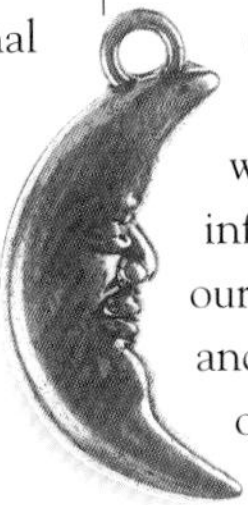

ABOVE AND RIGHT ***Day and night are the most obvious cycles.***

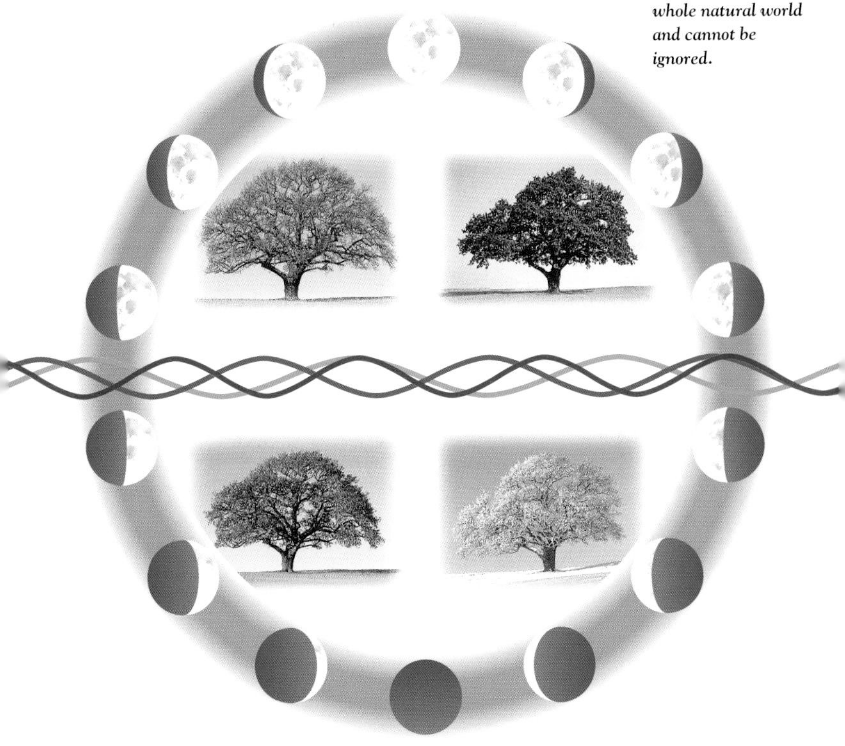

BELOW ***The seasons of the year affect the whole natural world and cannot be ignored.***

THE THREE MAJOR CYCLES

There are three major biorhythm cycles that each have a profound influence on our daily lives:

Physical
The first cycle affects our physical responses and lasts for 23 days.

Emotional
The second cycle governs our emotional outlook and lasts for 28 days.

Intellectual
The third cycle influences our intellectual perceptions and lasts for 33 days.

All three cycles run concurrently, of course, but independently and not necessarily out of phase with each other.

Each rhythm has three critical or change days that occur at the beginning of the cycle, at the halfway stage when the phase changes from positive to negative, and at the end, which is also the start of the next positive phase.

It is when a rhythm changes from one phase to another that we are vulnerable to difficulties. But when we calculate and chart these cycles a new and exciting world of predictable behavioral patterns appears.

RIGHT *We can feel very fit during the first half of the physical cycle.*

LEFT *Our position in the emotional cycle affects our relationships.*

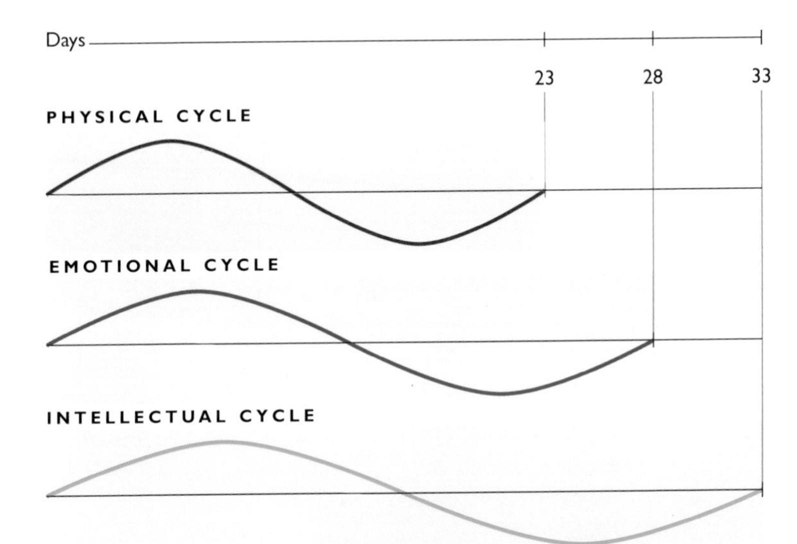

ABOVE *Our daily lives are governed by three major cycles: the physical, emotional, and intellectual cycles.*

BELOW *The beginning, middle, and end of each cycle are critical days when the phase changes from positive to negative – or from negative to positive.*

LEFT *At 33 days, the intellectual cycle is the longest of the the three.*

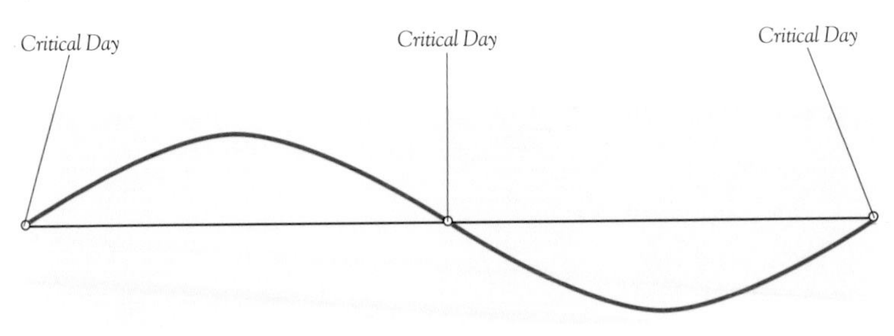

A short history

A HUNDRED YEARS AGO *a noted ear, nose, and throat specialist in Berlin, Wilhelm Fliess, observed unusual but regular patterns of behavior in two significant areas – physical and emotional. He observed the patterns, kept records, and discovered a physical cycle that lasted for 23 days. He then found an emotional rhythm that lasted for 28 days.*

ABOVE ***Wilhelm Fliess, one of the first to notice biorhythms.***

OTHER DISCOVERIES

At about the same time in Vienna Professor Hermann Swoboda, an eminent psychologist, independently came to similar conclusions. Later, in the 1920s Alfred Teltscher, a doctor of engineering and a mathematical wizard, discovered the intellectual cycle of 33 days. And at the same time, Rexford Hersey and Michael John Bennett, doctors at Pennsylvania University arrived at similar conclusions stemming from the results of their own independent research.

LEFT ***Hermann Swoboda was also a biorhythm pioneer.***

But for a long time the significance of these rhythms was ignored because no one could establish a simple method for calculating them. Since then a few dedicated people have barely managed to keep the study alive. However, calculations have now been simplified and these are dealt with later in this book. Other cycles that may affect our behavioral patterns have also been discovered, but none has stirred the same interest and excitement created when the original physical, emotional, and intellectual rhythms were discovered.

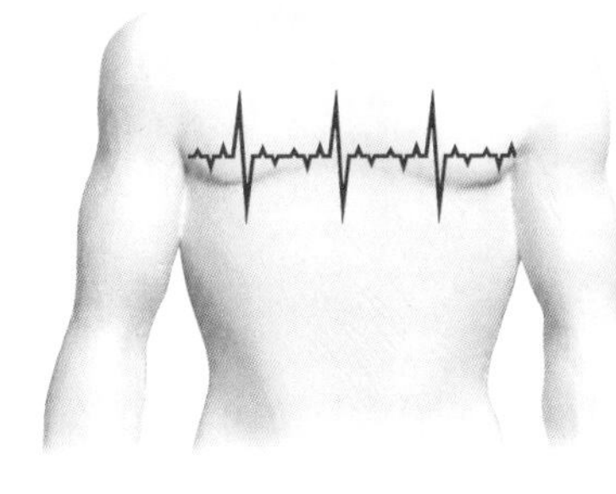

RIGHT ***Heartbeat is a significant force in our behavior.***

A student of biorhythms can predict when we are liable to be "off-balance" and make adjustments to help suit these particular moods. With this foreknowledge we can then steer carefully through our working day to suit the energy reserves we know we currently possess. This is better than going blind into a situation for, when we are prepared, we are best able to take control of our lives.

BIORHYTHMS AT WORK

One of the other cycles that affects us concerns our heart: this beats around 106,380 times in a normal day but, if our routine or health is disturbed, the degree of fluctuation of our heartbeat can affect the way we respond to people or situations in which we find ourselves. When we are startled we show fear, the heart beats faster, and then it takes time to return to normal once the danger is past. When we are nervous or overconfident our heart may not perform efficiently and this is the time when we may make mistakes. This is something of which we are all aware.

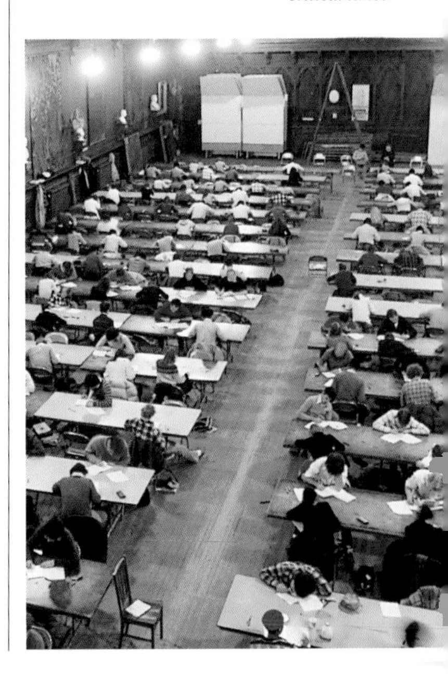

BELOW ***Biorhythmic awareness helps us prepare for critical times.***

The cycles at work

A CYCLE *is a recurrent round or period of events, of phenomena, or of time. The illustration shows a cycle as a sine wave. It starts with a high period, then sinks to a low stage, and the gap between the two extremes reveals its frequency or length of time. The height or amplitude of the wave shows its strength. If the wave is part of a series, the end is also the beginning of the next cycle.*

THE FIRST HALF

Each of these rhythms begins on the day we are born and continues throughout our lives until the day we die. The first half of a cycle is always the ascending or positive and progressive period. Capabilities and performance are usually at their best at this stage. Abilities remain strong until the cycle enters its second half.

RIGHT ***The cycles continue throughout the year.***

BELOW LEFT ***Each of our rhythms begins with our birth.***

RIGHT ***The first half of each cycle is active and positive.***

THE SECOND HALF

The second half of the biorhythm is the rejuvenation or recuperative phase when energies are at a low ebb. The body rests and prepares for the next positive phase.

RIGHT ***The second half of each cycle is a period of regeneration.***

BELOW RIGHT ***Our biorhythms continue throughout our lives.***

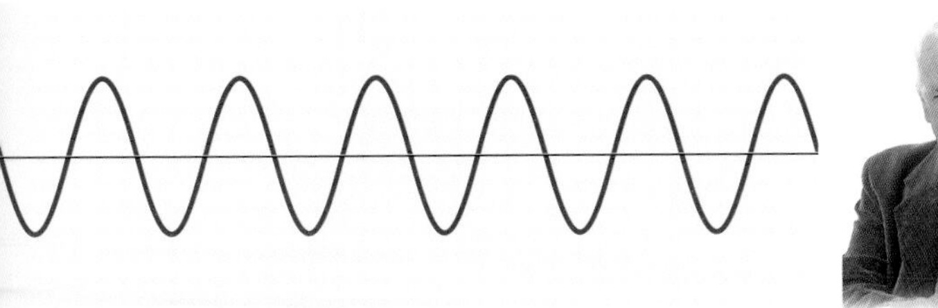

CRITICAL DAYS

The start, the halfway point, and the end/beginning of a cycle are always periods of caution and are known as critical days. Usually, nothing critical actually happens except for the phase change itself.

You can compare critical days to the moment when a light bulb is switched on or off. When you flick the switch the bulb is suddenly filled with power and energy. It transforms from a nil-energy, negative phase to a glowing and positive stage. This is when bulbs usually blow. A bulb may also blow when the light is switched off and power no longer flows through it – it switches from positive to negative. It fails because after a time it is weakened by the sudden energy fluctuation or temperature change.

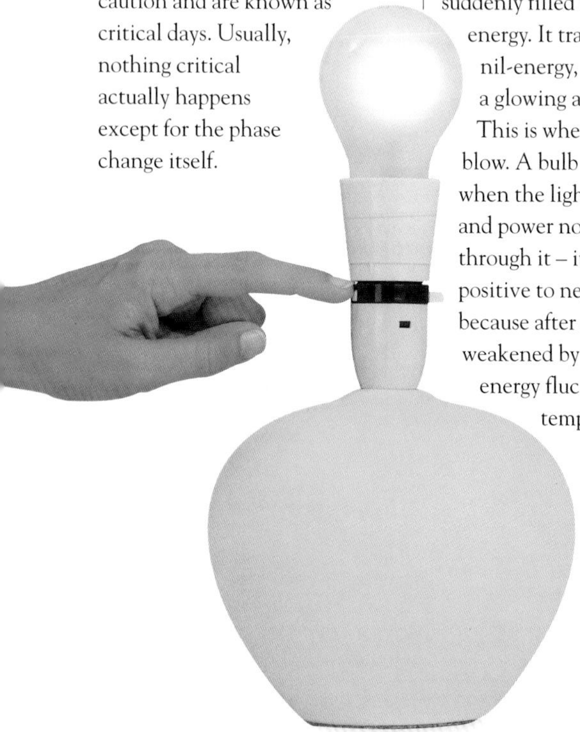

LEFT ***The phase change in a critical day is as sudden as a light turned on or off.***

BELOW ***The critical days in each cycle.***

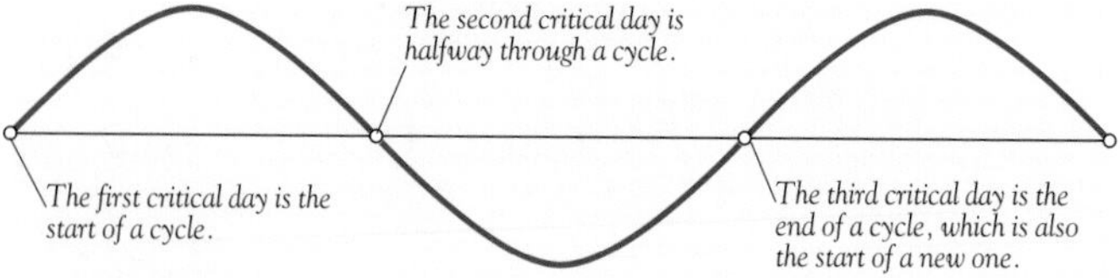

DAYS OF CHANGE

And so it is with people. During changeover periods we become more vulnerable, we are temporarily off-balance and our normal reactions slow down: we become accident-prone. During the first, or positive, phase of a cycle we feel much more alert, emotionally responsive, and perceptive.

In the second half, or negative stage of the cycle, we become less responsive, less alert, and less perceptive. On a critical day, when the cycle changes polarity, we are at our most vulnerable. Our judgment becomes faulty or unreliable, accidents can happen, and we should defer important or strenuous activities since we may feel less cooperative and liable to make poor decisions. We should try to avoid physically overdoing things and not allow small issues to get us down. If possible, delay making important decisions until after changeover periods.

BELOW ***On the midway point of a cycle we are prone to be clumsy.***

LEFT ***Our judgment is better on days that are not changeover days of a cycle.***

The physical cycle of 23 days

THE FIRST STAGE *of the physical cycle pumps you full of energy and you feel good and on top of the world. Nothing is too much of an effort and you feel able to achieve your maximum potential.*

POSITIVE PHASE

Feel free to participate in most physical activities without fear of overstretching yourself, but take care not to overdo things. Day seven, in the middle of this first phase, is a mini-critical day when the cycle peaks. At this point you are at your best in the physical sense and able to cope with everything that comes your way. Make the most of this and plan your time and activities accordingly.

ABOVE ***Day seven is the peak of the physical cycle.***

BELOW ***Physical ability follows a set pattern.***

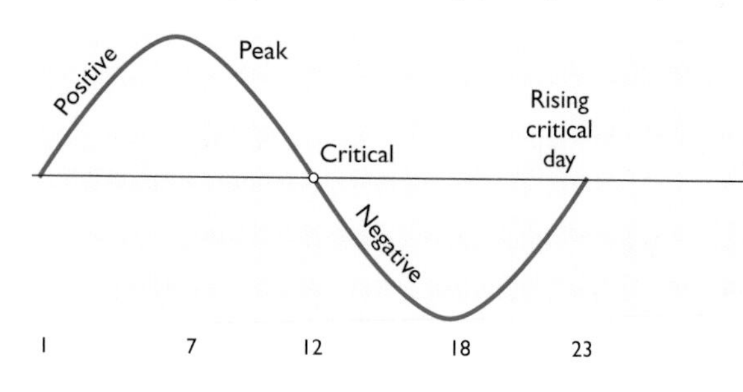

NEGATIVE PHASE

After this peak, the cycle turns downhill and passes through the baseline on day 12 to the negative phase. This is a critical day, a day of caution. You may easily misjudge matters, feel slightly off-balance, and your reactions may be noticeably slower – you are accident-prone!

The second phase continues downward and troughs on a second mini-critical point at the 18th day, the lowest point of the negative stage. Conserve your energy and take things steadily, one step at a time.

ABOVE *You can find yourself a little off balance on a critical day.*

RIGHT *Take things gently at the bottom of the physical cycle.*

MOVING UPWARD

Then the cycle begins to move upward, slowly gathering more energy and crossing the baseline on day 23, the second critical day. You will feel buoyant and positive again, but take care: don't be too confident – accidents can happen! Some people actually feel a surge of physical energy on a positive or rising critical day.

CAUTION

Some people have experienced a brief period of lethargy during the upward move toward the end of the cycle. This may take the form of unexplained fatigue; it's best simply to go with it – let go and relax in an armchair for an hour or two until the feeling passes.

The emotional cycle of 28 days

THE EMOTIONAL CYCLE *is largely concerned with your mood and social ability. In the first stage of this rhythm you are cheerful and cooperative, optimistic, you shine socially, and you are highly creative in your outlook.*

ABOVE ***Social feelings peak on day eight.***

BELOW ***The stages of the 28-day emotional cycle.***

On the eighth day, in the middle of this first stage, lies the mini-critical period when the cycle peaks. You are now at your most amenable, socially and emotionally. After this peak the rhythm turns downward and, on the 15th day, the critical day crosses the line into the negative stage. You are likely to tense up, be irritable or ultrasensitive, and easily stressed. Accidents are more likely to happen on this day.

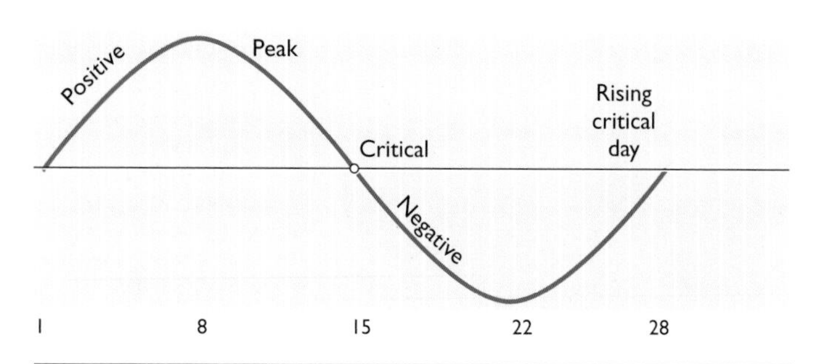

NEGATIVE PHASE

The negative phase continues down and troughs at a second mini-critical point on day 22, the lowest point of the negative cycle. You may feel left out of things, vulnerable, and raw nerved. This is your worst point for contact with others. This short period would be best spent on your own. If this is impossible, try to avoid getting into confrontational situations.

RIGHT ***Sensitivity to criticism diminishes as the cycle ends.***

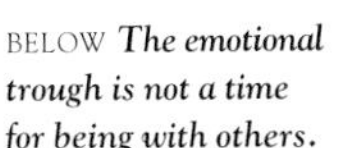

BELOW ***The emotional trough is not a time for being with others.***

MOVING UPWARD

The cycle then begins to move upward; slowly you gather enthusiasm for social and emotional commitments and life in general. The cycle passes through the baseline on the 28th day – the second critical day. Here you may overreact to criticism and be oversensitive to anyone or anything that does not perform properly. Thinking may be clouded by your emotional state. The heart rules the head. It is easy to remember this caution day because it occurs every other week on the same day: it has a continuous cycle of exactly 14 days.

The intellectual cycle of 33 days

THE INTELLECTUAL CYCLE *governs your reasoning, perception, and intellectual abilities. Logic and mental ability are at their height in the positive phase. Ideas flow well, problems are easily resolved, and study periods are profitable.*

POSITIVE PHASE

This cycle peaks on the ninth day, a mini-critical day, at which point your intelligence is at its best. You are at your most perceptive, your mind is agile, and it is easy for you to deal effectively with several issues at any one time. This is a good time to take on something completely new.

BELOW ***The stages of the 33-day intellectual cycle.***

RIGHT ***On day nine your brain is at its most efficient.***

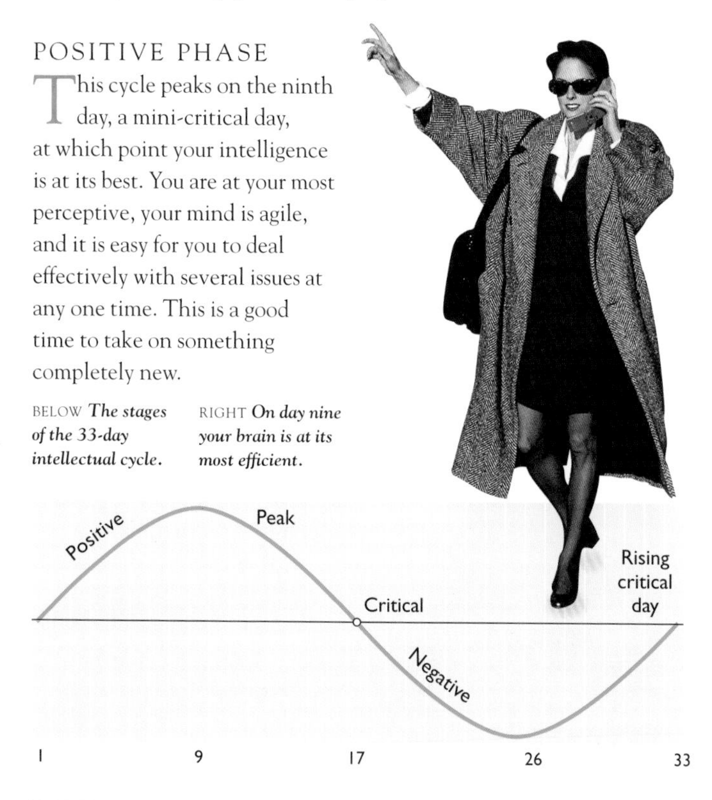

RIGHT *In the negative phase you may not express yourself well.*

NEGATIVE PHASE

After this peak the rhythm turns downward and, on the 17th or 18th day, the cycle crosses the baseline into the negative phase and the critical day. Now you are liable to make bad decisions and may be absent-minded. You may have difficulty expressing your thoughts, and errors could occur through simple misunderstanding. The negative phase descends until it troughs on the second mini-critical point at day 26, the lowest point. Here your memory may become muddled, your common sense faulty, and you find yourself making silly mistakes.

MOVING UPWARD

After this, the cycle begins to rise upward and the mind becomes less clouded. It is an ideal time for reappraisal or studying old material, but not yet the moment to take anything new on board. As the cycle passes through the baseline you come to the second critical period on the 33rd day. This cycle is longer than the physical and emotional cycles where critical periods last for 24 hours. Allow 48 hours for the critical "days" in the intellectual cycle and do not tackle matters that need a clear mental approach until these periods have passed.

BELOW *The upward swing of the intellectual cycle is a good time for reappraisal.*

Biorhythmic calculations

THIS STEP-BY-STEP GUIDE *will enable you to calculate biorhythms for any date with a simple pocket calculator. Once you have established the figures for each of the cycles, it is easy to chart your own biorhythms, or those of someone else.*

EXPERIMENT WITH CHARTS

Using biorhythm charts helps you to plan your own life more effectively, and it may also help you to understand other people's unexplained changes of mood or unexpected actions. You may be curious about a friend's strange behavior on a certain date, for instance, or you may wish to set up an advance chart ready for a specific activity.

LEFT ***Even great geniuses, such as Albert Einstein, are affected by their biorhythms.***

Provided you have the birthdates of the people who interest you, plotting biorhythm charts is a simple matter.

ABOVE ***All you need to chart someone's biorhythms is his or her birthdate and a calculator.***

Students of history can create charts for the heroes and villains of the past and gain insight into their actions on the dates in question. At the back of this book is a page of blank charts to make copies and experiment with, and the biorhythmic charts of some famous figures of history and of our own times who have come to tragic ends are discussed on pages 54 to 57, to show how an understanding of their biorhythms can add to our understanding of events.

HOW TO CALCULATE BIORHYTHMS

You can use the same basic formula given below to calculate the biorhythms for any person, living or dead, at any moment in his or her life, including the future.

Calculate your base figures for the first day of a month for this allows you to work backward or forward with reasonable ease. Use the tables in the column (right) to help you find the total number of days easily and combine the necessary figures to arrive at the total.

- Start by using the birthdate of the person in question (day, month, and year), to work out the exact number of days from their birth to the day in which you are interested. Add in the extra days for leap years and months, bearing in mind the different lengths of months.
- To calculate the physical cycle, divide this number by 23.
- To calculate the emotional cycle, divide this number by 28.
- To calculate the intellectual cycle, divide this number by 33.
- In each calculation there will be a remainder of several decimal points.
- To find out the biorhythmic stage of the particular date in which you are interested, treat these remainders as whole numbers and divide each of them by 23, 28, and 33 respectively. Round the numbers up to the nearest whole number. This will give the biorhythm stages for the date in question, when calculated to the nearest whole figure. (See overleaf for a sample calculation.)

YEAR CALCULATIONS

This table enables you to see at a glance how many days there are in any combination of years up to 90. Use it to make the base calculation according to the age of the person whose biorhythms you are charting.

1 x 365 = 365	10 x 365 = 3650
2 x 365 = 730	20 x 365 = 7300
3 x 365 = 1095	30 x 365 = 10950
4 x 365 = 1460	40 x 365 = 14600
5 x 365 = 1825	50 x 365 = 18250
6 x 365 = 2190	60 x 365 = 21900
7 x 365 = 2555	70 x 365 = 25550
8 x 365 = 2920	80 x 365 = 29200
9 x 365 = 3285	90 x 365 = 32850

LEAP YEARS SINCE 1752

You need to allow for the extra day in every leap year, which occur every four years. The table below lists the leap years from the mid-18th century onward so that you can build them into your calculations.

1756	1760	1764	1768	1772	1776
1780	1784	1788	1792	1796	1804
1808	1812	1816	1820	1824	1828
1832	1836	1840	1844	1848	1852
1856	1860	1864	1868	1872	1876
1880	1884	1888	1892	1896	1904
1908	1912	1916	1920	1924	1928
1932	1936	1940	1944	1948	1952
1956	1960	1964	1968	1972	1976
1980	1984	1988	1992	1996	2000

EXAMPLE

Take someone born on October 2, 1977 and calculate the cycles for December 21, 1999 to see what sort of Christmas that person can expect.

1 On December 21, 1999, our subject is 22 years old plus the extra days. Thus:

20 years = 7,300
+ 2 years =730
+ extra days = 80
+ leap years = 5
+ one day = 1
= Total 8,116 days since birthdate.

2 Use your calculator to divide this number by 23 (for physical), then 28 (for emotional), then 33 (for intellectual).

8,116 ÷ 23 = 352.86956
(Physical cycle)

8,116 ÷ 28 = 289.85714
(Emotional cycle)

8,116 ÷ 33 = 245.93939
(Intellectual cycle)

3 Using a calculator, multiply each of the remainders by the cycle numbers. This will give the biorhythm stage of the cycle for the day in question.

Thus:
.86956 x 23 = 19.99988
– which, to the nearest whole number,
is 20 for the physical cycle.

.85714 x 28 = 23.99992 –
which, to the nearest whole number,
is 24 for the emotional cycle.

.93939 x 33 = 30.99987 –
which, to the nearest whole number,
is 31 for the intellectual cycle.

LEFT *In making calculations, round decimals up or down to the nearest whole number.*

SUMMARY

Christmas for our subject looks bleak with critical days on the 24th, 25th, and 26th but the New Year will go well because all the rhythms are in positive mode. This is shown as "PPP" on page 31.

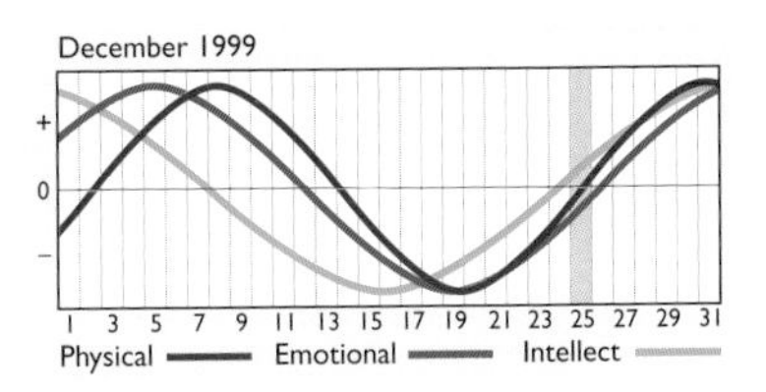

ABOVE *This chart shows what kind of Christmas vacation its subject can expect.*

MAKING A BIORHYTHM CHART

To draw up a chart take each biorythm stage in turn.

For the physical cycle of our subject the biorhythm stage of 20 indicates the 20th day of the 23-day cycle. This cycle reaches critical point on the 12th day and again on the 1st day of the next cycle (see pp. 16–17) so that, on the 20th day of the cycle, our subject is 4 days away from a critical point in the second or negative phase.

Our calculations are for the 21st day of the month, therefore, the next critical stage is on the 25th December.

Treat the Emotional and Intellectual cycles in the same way (see pp. 18–21), counting forwards and backwards to the critical points, from the day in question, and draw in the curved line using a protractor and a different colored pen for each cycle. The same method applies when drawing up a chart to the 1st day of the month.

LEFT *You can plan activities to match your biorhythmic stages.*

Compatibility

HOW COMPATIBLE ARE *you with your partner? Perhaps out-of-phase biorhythms are the cause of those black days when the pair of you inhabit different worlds.*

RIGHT ***You can assess your compatibility with your partner by comparing charts.***

YOU AND YOUR PARTNER

To assess biorhythmic compatibility, the difference between your rhythm and your partner's rhythm is calculated in days. Each day represents a proportional difference in the cycles and the percentage figures are always constant. The easiest way to calculate this is to see how many days apart your and your partner's critical days appear.

COMPATIBILITY EXAMPLES

If you and your partner are four days apart in your physical cycle, the table shows that you are 65% compatible. If you are seven days apart in the emotional cycle, the table confirms 50% compatibility. A twelve-day difference in the intellectual rhythm shows you are 27% compatible. The overall assessment is only a guide; the individual comparison figures are the most important.

COMPATIBILITY TABLE

DAYS APART IN CYCLE	PHYSICAL CYCLE %	EMOTIONAL CYCLE %	INTELLECTUAL CYCLE %
0	100	100	100
1	91	93	94
2	83	86	88
3	74	79	82
4	65	71	76
5	57	64	70
6	48	57	64
7	39	50	58
8	30	43	52
9	22	36	46
10	13	29	39
11	4	21	33
12	4	14	27
13	13	7	21
14	22	0	15
15	30	7	9
16	39	14	3
17	48	21	3
18	57	29	9
19	65	36	15
20	74	43	21
21	83	50	27
22	91	57	33
23	100	64	39
24	71	46	
25	79	52	
26	86	58	
27	93	64	
28	100	70	
29	76		
30	82		
31	88		
32	94		
33	100		

Add the results and divide by three to arrive at the overall compatibility figure.

PHYSICAL COMPATIBILITY

An 85% to 100% compatibility is fine for most physical activities that require joint effort; 75% requires the temporarily stronger person to take the initiative, but he or she must allow for the partner's inability to be as active.

Less than 50% means that both partners must exercise good judgment to time events for their mutual satisfaction.

ABOVE ***Marriage needs good emotional compatibility.***

RIGHT ***Less than 50% emotional compatibility calls for careful judgment.***

EMOTIONAL COMPATIBILITY

A 75% to 100% compatibility is good. Long-term associations like marriage can lead to tension because similar rhythms lack stimulus. A 45% to 65% compatibility is best for most married couples. Less than 35% is unfavorable. Partners do have to exercise a lot of tact to maintain emotional rapport.

A high proportion of married couples or people who have lived quite happily together for some time frequently yield quite low compatibility ratings.

INTELLECTUAL COMPATIBILITY

A 100% compatibility is good but, because partners are always in the same phase, the lack of mental stimulation creates boredom. A percentage around 65% to 75% is ideal: people tend to work better together.

Under 50% means that partners must monitor themselves all the time, which can be tiring. A high intellectual rating between an employer and employee is nearly always found in successful business partnerships.

LEFT ***Shared interests result from intellectual compatibility.***

OVERALL COMPATIBILITY

The higher the overall assessment percentage, the better the basis for any relationship. The highest individual cycle rating usually reflects the strongest element in a relationship. For instance, a high physical figure implies strong physical attraction; a high sensitivity rating reflects a deep emotional rapport; and, when the intellectual percentage is the highest, the association depends on the strength of the mental attraction between partners.

Do a compatibility check on a group of people such as your family, workmates, team members if you play a sport, or college friends. See if it confirms your observations about the relationships within that group.

ABOVE ***About 45% to 65% is the optimal level of compatibility for married couples.***

The daily combinations

EACH BIORHYTHM INDIVIDUALLY *experiences the three continuous stages daily. They are either in the plus phase, the negative stage, or are critical. And when they are critical, they may change from positive to negative, or negative to positive.*

BIORHYTHMS IN ACTION

The interaction between the biorhythms creates 27 different combinations that may be experienced at any one time. Assessments of the different combinations are set out in the next few pages and show comparisons for the 10th day of the month.

However, these are meant as guidelines only. We all differ in the way we react to any given set of circumstances. Personality, health, age, and upbringing all play a part in defining our mood, but calculating biorhythms as well can help alert us to possible problems before they occur.

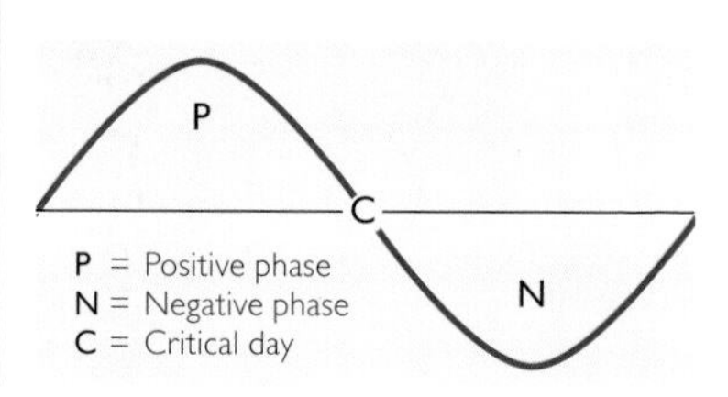

THE 27 COMBINATIONS

NNN When all three cycles are in the negative phase you will be below par. Relax. Don't start anything new, for even special efforts may not last.

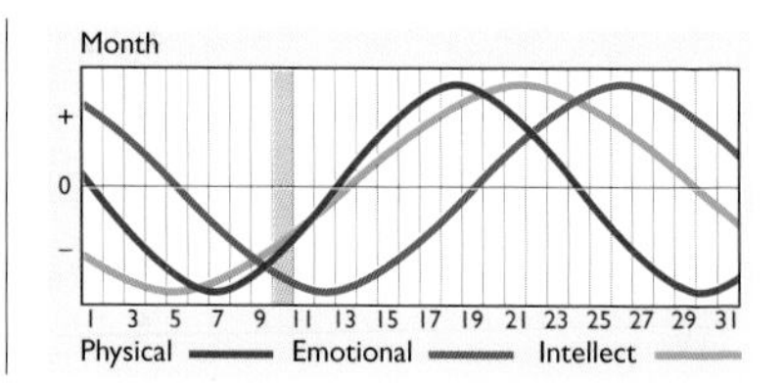

PNN Physically you feel fit, but it would be better to leave social activity or studying for another time. Think before you commit yourself to anything new at this time.

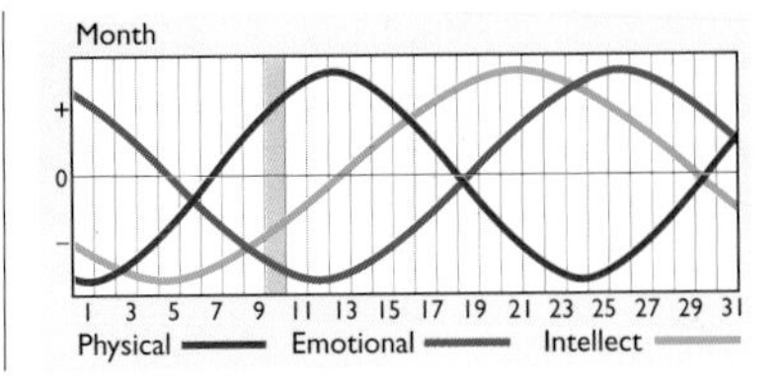

PPN Emotionally and physically you feel good, but your judgment may be suspect. Fairly positive for the most part, let instinct be your guide – work on first impressions.

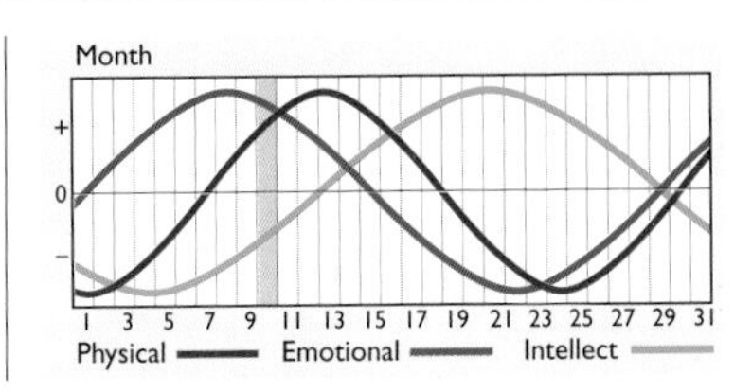

PPP This is quite clearly a good day all round. Give full rein to all you do and goals should be achieved quite easily. You might feel a little restless if you have nothing planned.

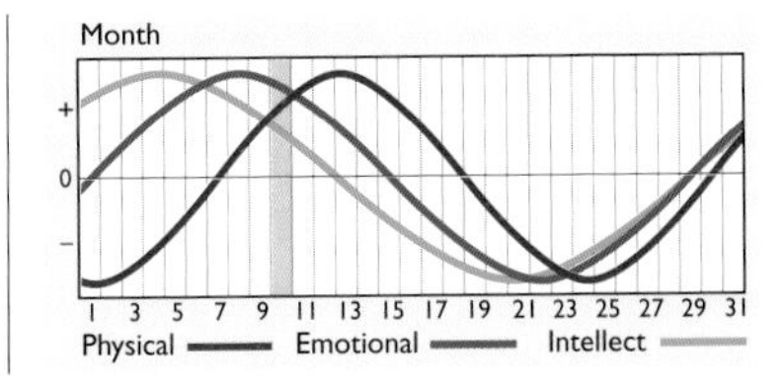

NPP The spirit is willing, the body isn't! Could be a problem keeping up with your emotional and intellectual pursuits without getting tired. Plan carefully; important things first.

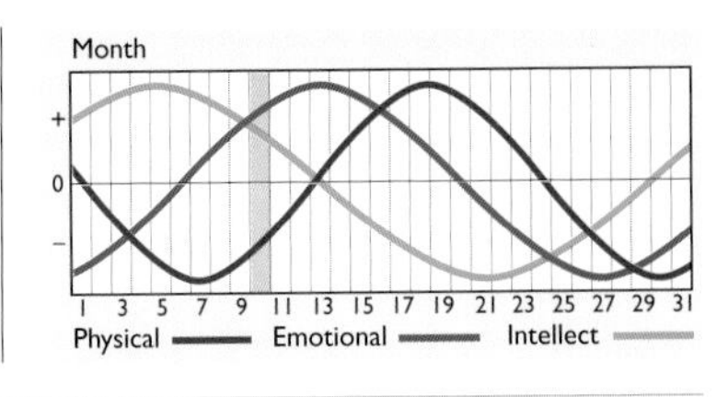

NNP Your mind is likely to take control of everything. Emotionally, you're not too sociable and might lose your temper; you won't suffer fools today so avoid confrontations.

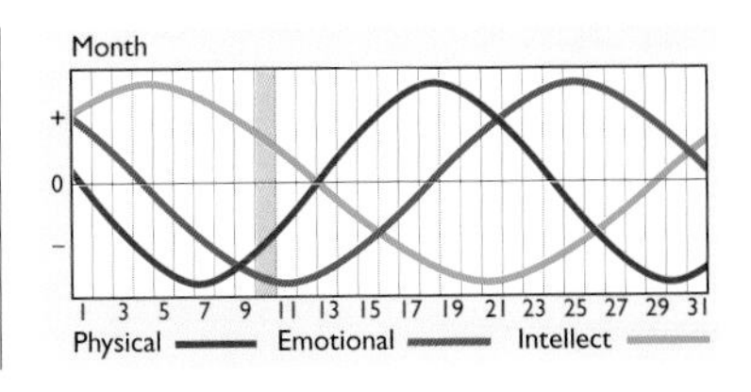

LEFT *There are days when you should avoid delicate tasks.*

NPN You are well disposed to most people, but energy levels are too low to allow much concentration. Creative types should spend time with their hobbies or pastimes.

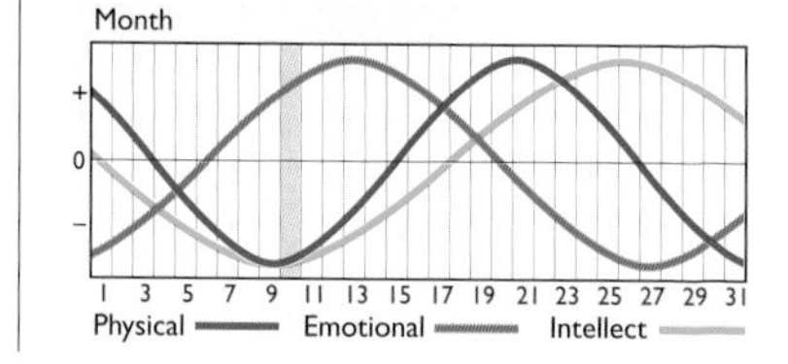

PNP You may feel a little down. Physically you should be fine, with intellectual ability well above average. However, it might be an idea to avoid social gatherings.

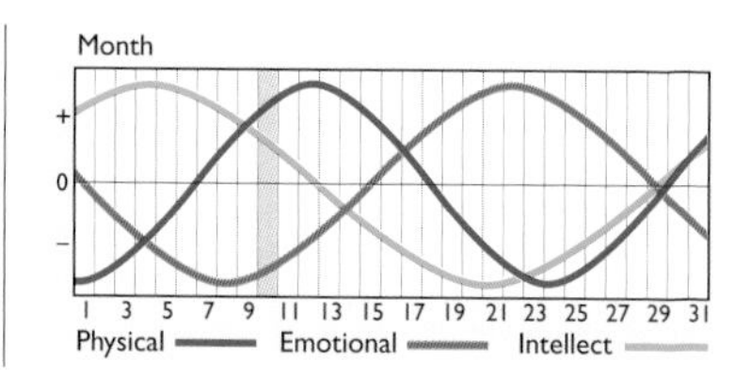

NNC You are accident-prone. Physically you will tire easily and you are not too sociable. You may be unable to think clearly, so try not to take short cuts or invite risks.

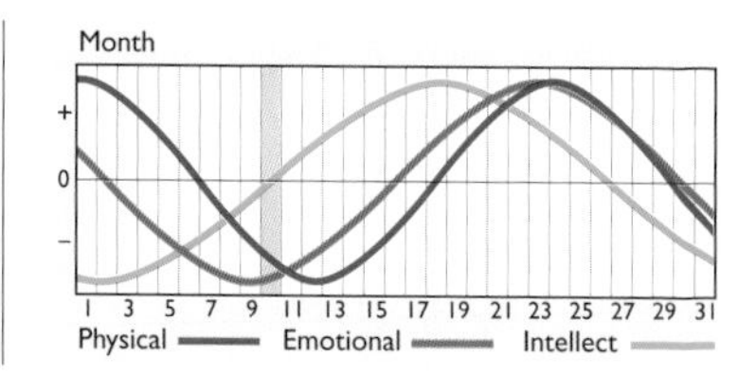

NPC Another accident-prone day so keep a low profile. Physically you are low, but intellectually critical. Only your emotional outlook seems to function properly.

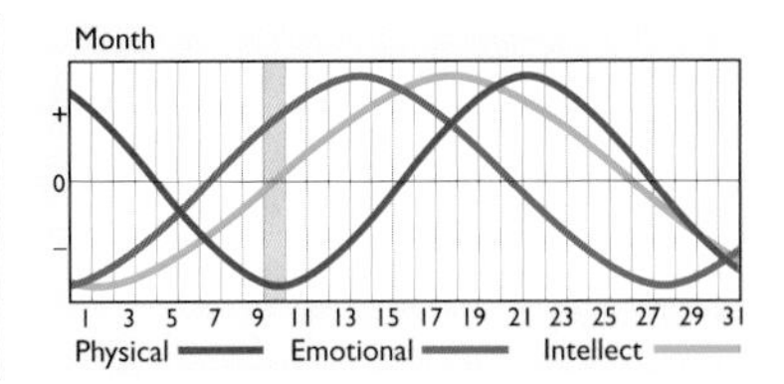

PNC It may be difficult to think straight. You feel unsociable and restless physically. You could also feel rather unsettled. This is another accident-prone day.

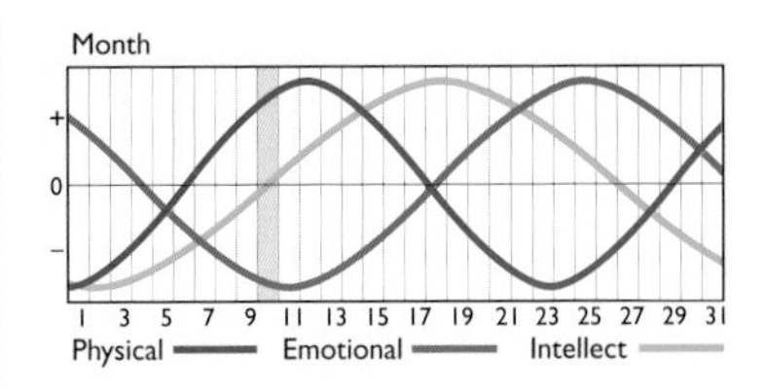

PPC Both physically and emotionally you are in fine fettle. You can be a trifle overconfident and liable to make small errors of judgment because you are also off-balance intellectually.

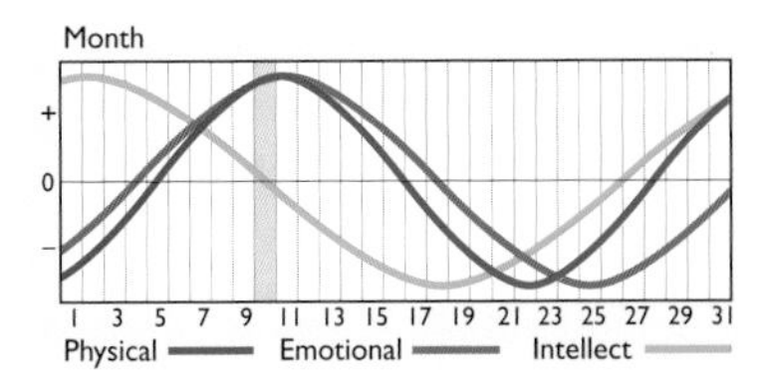

NCN Hasty decisions now mean problems later. You are intellectually and physically down with an emotionally critical day. Be careful: it can lead to accidents.

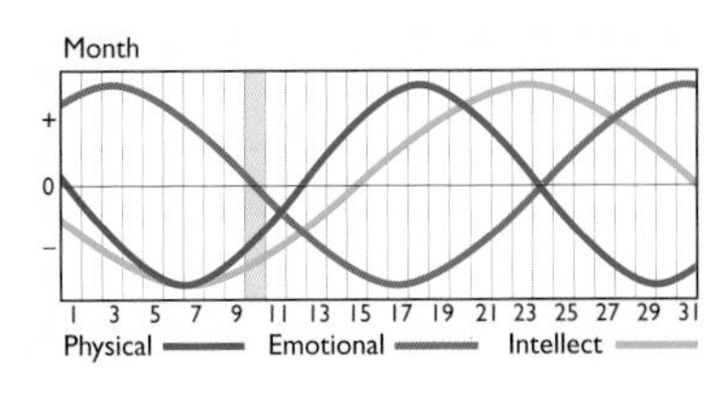

NCP If driving or using unfamiliar tools and equipment, you may make elementary mistakes and have or cause an accident. You are sensitive to criticism, no matter how well meant.

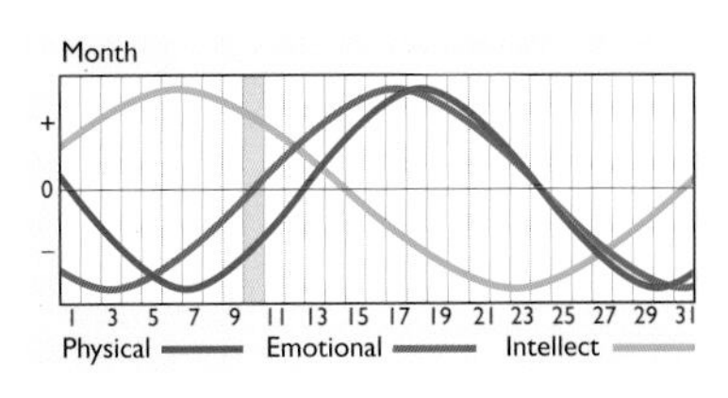

PCN Physically you are in top form but at odds with everyone else because you fail to grasp points. Not exactly sociable either. Take care when studying or revising notes.

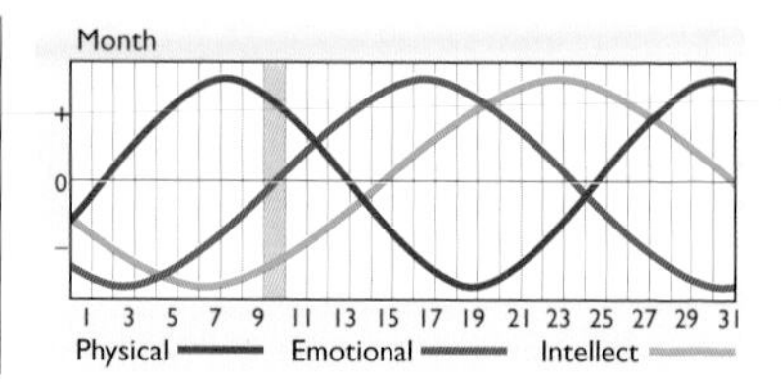

PCP Irritable with a side salad of over-confidence. You may feel the need to achieve something, but a lack of concentration makes this difficult. Don't take on extra work.

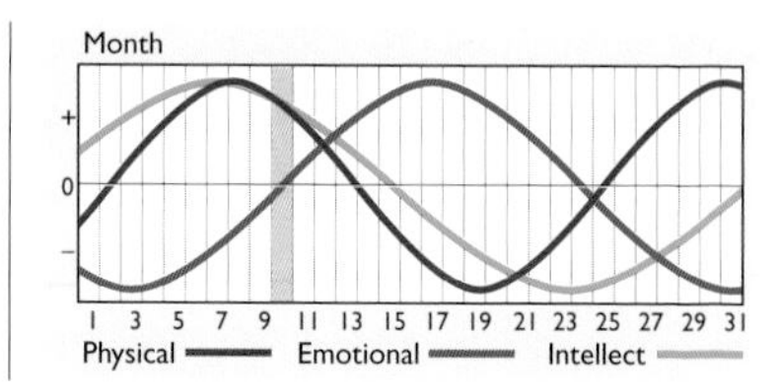

CNN This is a short and rather serious accident-prone time. Do take care. You are off-balance physically and not as perceptive as normal. Be careful at all times for your safety.

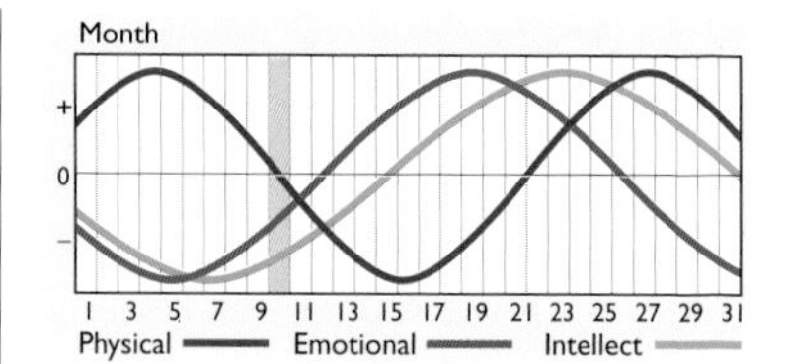

BE AWARE OF YOUR BIORHYTHMS

If you know your cycles are adverse for a planned activity it makes sense to avoid that activity until a more propitious time is indicated.

RIGHT *Avoid potentially dangerous activities on accident-prone days.*

CNP You may tire easily and, since you feel low emotionally, you are inclined to give social activities a miss. Keep your mind clear and remember, distractions lead to accidents.

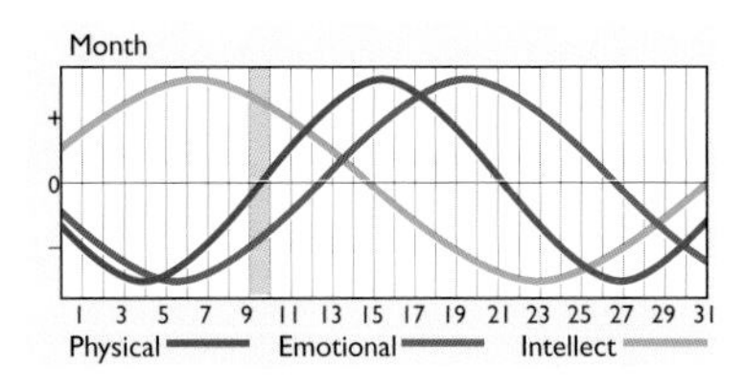

CPP An emotionally and intellectually high day but physically overtaxing. You think you can do it all, but you can't! You are accident-prone at this time, so don't forget it.

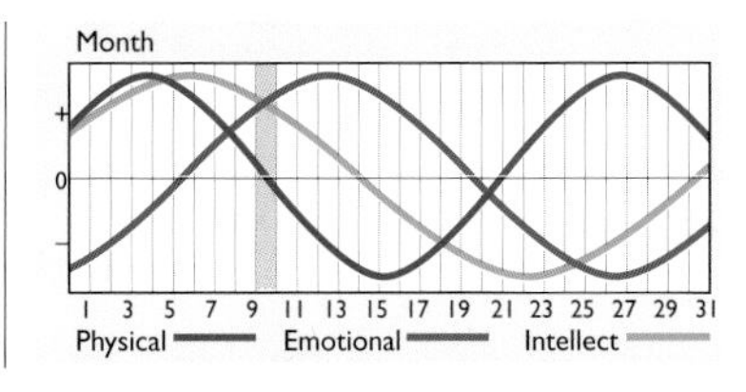

CPN Far from your best, really. Your perception is high, but you feel a little sluggish physically. Coordination might be a problem and could lead to minor accidents.

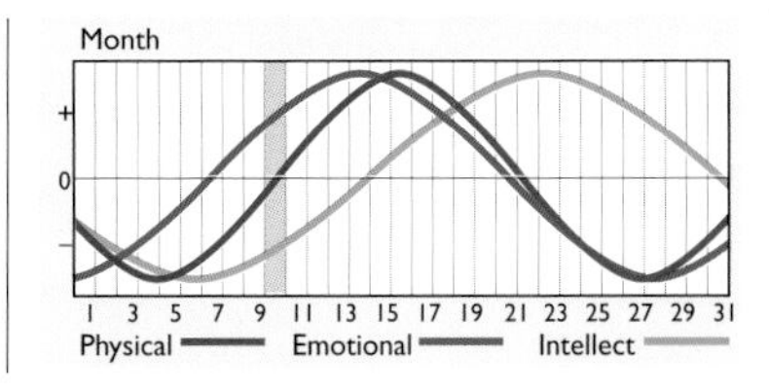

NCC A poor day with your emotional and intellectual rhythms well out of kilter. With your physical rhythm in negative phase it might just make things worse. Avoid physical activity.

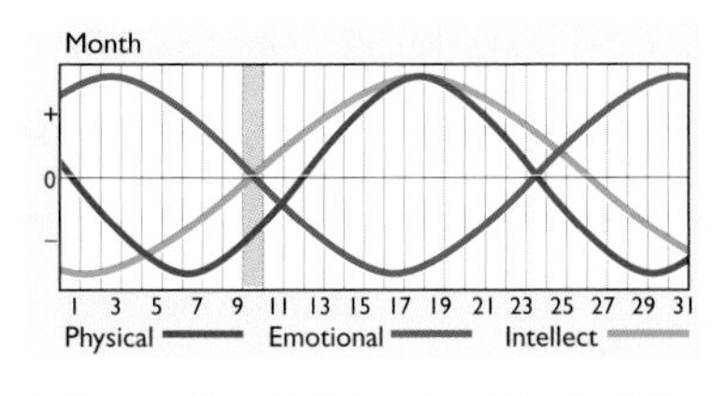

PCC A high-risk day again. You are prone to talk yourself in and out of problems because you are off-balance, probably more than you realize. Physically, you are in top form.

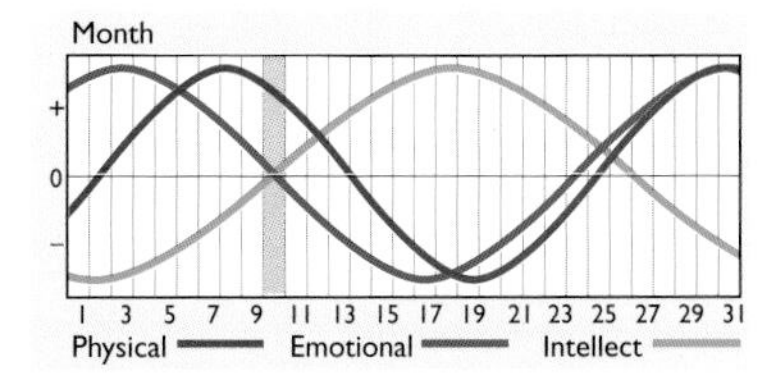

CNC An irritating period with everyone thoroughly uncooperative – or at least that is how it seems. A low emotional rhythm and two critical days. Accidents may happen.

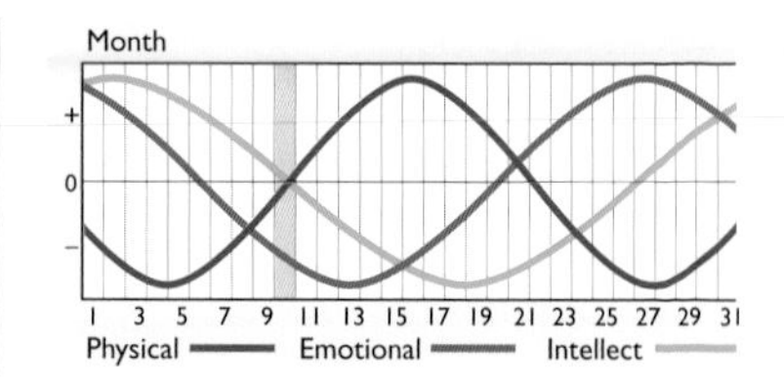

CPC You will feel off-balance, with both physical and intellectual cycles in their critical stages. However, the positive emotional rhythm eases matters a little.

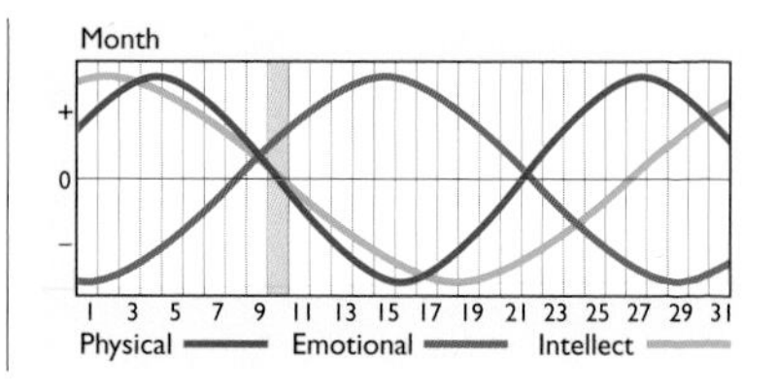

CCP On emotional and physical critical days there is always a danger of a serious mistake happening when you least expect it. Awkward day – think things through first!

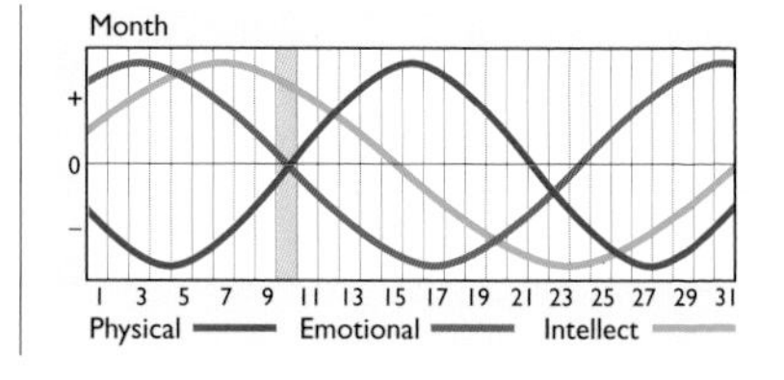

CCN Perception and understanding are low. With both physical and emotional cycles at a critical stage, you are unlikely to see a problem until it is too late to do anything about it.

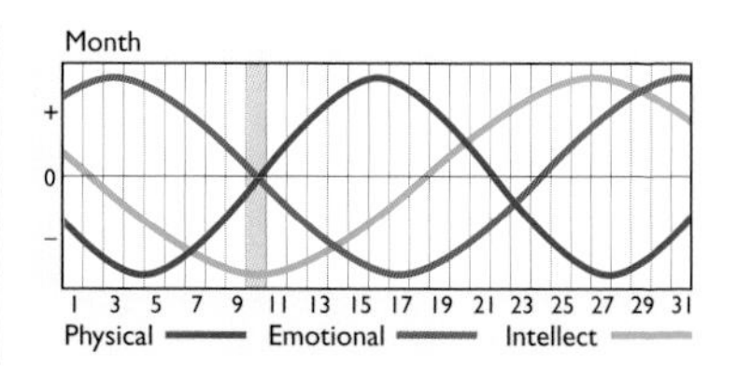

CCC Take exceptional care. Three criticals at once does not happen very often but, when it does, you must make every effort to remain cool, calm, and very collected.

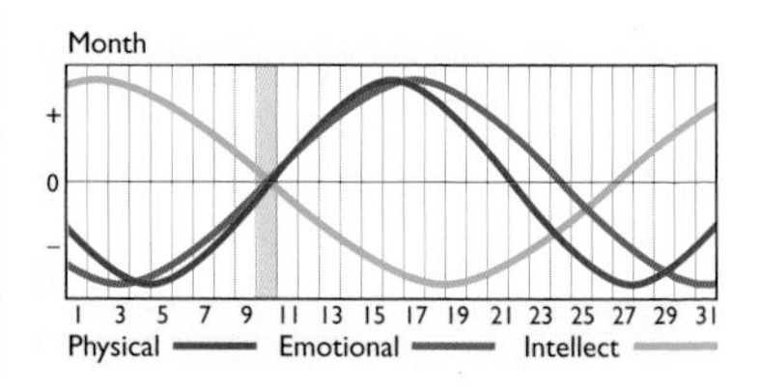

Planning ahead with biorhythms

ONCE YOU ARE COMFORTABLE *with the charting of biorhythms, you will start to find an increasing number of uses for them in your daily life.*

PREPARING FOR THE FUTURE

Bear in mind that these combinations are only a guide, not a statement of fact. Life is different for all of us, and foretelling the future is a notoriously difficult science. Some of these conditions last only for a day, but there are times when certain combinations may persist for six or even seven days. This happens more with the positive or easy-going phases than the negative ones.

PLAN WITH YOUR RHYTHMS

Because of this, take time to chart your personal biorhythms for the next six or twelve months and this will put you more in control of your own destiny. Learn to prepare for the days where attention to detail can make the difference between success and failure. No one likes a loser – but we all envy a winner!

To capitalize on your natural energies, learn to plan ahead with the aid of your diary. When any awkward or potentially difficult times loom in the distance, and your biorhythms seem also to be adverse, postpone or bring forward the activity to better suit your abilities for that period.

This is proper, judicial planning that all successful people use for their personal benefit and profit.

ABOVE ***It is worth taking time to plot your biorhythms ahead.***

Diet with biorhythms

LOSING WEIGHT IS *something most of us have attempted, but sudden crash diets, largely unplanned and unwelcome, tend to fall by the wayside within a few days. And many people feel guiltier about the failure of the program than about the reason for embarking on it in the first place.*

RIGHT ***Eating a sensible diet is difficult for many of us.***

CHART YOUR CRITICAL DAYS

With a little careful planning, using your biorhythms, there is no reason why you cannot implement and stick with a sensible regime to help you lose weight. Make up a biorhythm chart for two or three months in advance and highlight the emotional and intellectual critical days. Plan the diet to start about seven or eight days before the first of the critical days falls due and, if possible, in the middle of the physical positive phase.

Try not to wait too long for the ideal conditions or you may never get started. Lasting weight-loss can only really be achieved through long-term planning, in which the overall intake of high-fat and sugary foods is gradually decreased and a regime of healthy eating substituted instead. Plan to eat little and often – five or six times a day to start with – so that you avoid hunger-pangs and a craving for "quick-fix", sugary foods. Aim to eat five portions of fresh fruit or vegetables every day.

ABOVE *Sweet foods should be avoided on critical days.*

FORGET THE COOKIES

Careful eating is essential on physical critical days. Make sure you eat only healthy diet-conscious meals; forget high-sugar foods like donuts, cookies, or chocolate. If you must nibble between meals, then try a lettuce leaf or a raw carrot, but it should really only be necessary on a critical day.

RIGHT *If you have to have a snack, make it one of raw vegetables.*

Strengthen your resolve as each changeover period comes around. Monitor what you eat, but stick only with what you know will do you good. Watch when and what you drink in this period, particularly alcohol as this will increase the pounds. Drinking late in the evening may interfere with your regular sleep pattern. Resist the temptation to binge, and, as the days progress, so will your feeling of well-being. Working with your biorhythms in this way will give you a successful dietary system that does work!

Learn to study with biorhythms

EFFICIENT STUDY BRINGS *its own rewards, but it is not always easy to knuckle down to learn or study. Crash courses seldom work and can lead to poor health, especially if you stay up half the night on a physical low period or on any critical day.*

LEFT ***The negative intellectual phase is good for review or study.***

THE BEST TIME TO STUDY

However, it is when the intellectual cycle is in its negative stage that most people study the best. This is the prime review period. The mind does not want to absorb new data and is happy to run in neutral for a short while. Students find their minds are less perceptive. They can still learn about new things, but not with the same facility as when the intellectual cycle is in the positive phase.

Study and review periods are best when you are fit and in a high physical phase – with lots of stamina for long sessions. An active or plus intellectual phase is best for absorbing new ideas and taking in new information.

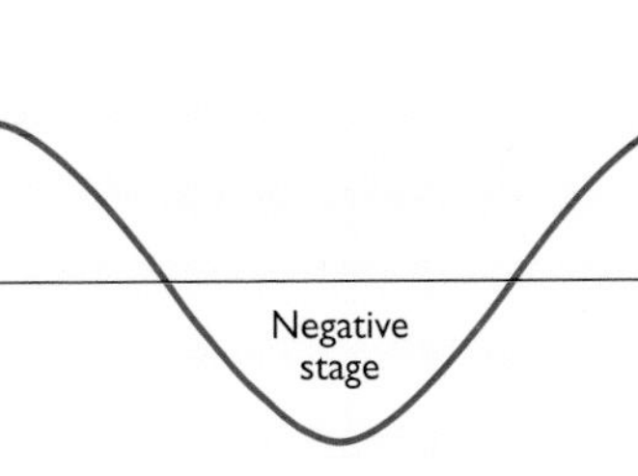

ABOVE ***The negative phase suits some tasks well.***

ABOVE ***Actors learn their lines best in the negative intellectual phase.***

USE THE NEGATIVE STAGE

Actors and actresses study and learn their lines and new roles best if the intellectual cycle is in the negative stage. Police officers have found this a good time to dig out old files and review case progress, and detection rates peak more in this period than at any other time. Conversely, sportsmen and -women may do less well.

LEFT ***Athletes do less well in the negative part of the intellectual cycle.***

Though normally relying on physical stamina more than intellectual prowess, they still need to keep their wits about them. Top athletes have been known to lose because they were in the negative intellectual phase and did not know it!

So, remember, when the intellectual rhythm is positive, you are better suited to taking on something completely new. However, you review old material best when this cycle is in the negative stage. Your brain is more receptive to a quiet revision of what it already knows.

Stop smoking with biorhythms

SMOKING IS AN ADDICTION, *so you need to plan effectively if you wish to give it up. You must have a well thought-out campaign to free the system of nicotine and you must really want to stop. With the help of biorhythms you can.*

CREATE A CHART

Make up your chart for at least one month, or two if you think it necessary. The best time to give up is at least two days after a positive to negative critical day in the emotional cycle. This allows you a period of nearly two weeks before the next critical day in your cycle when you will be feeling most vulnerable and unstable, and are therefore most likely to succumb to temptation.

Month
+
0
–
1 3 5 7 9 11 13 15 17 19 21 23 25 27 29 31
Physical Emotional Intellect

ABOVE ***Find the best days to start from your chart.***

Although it's not as important as the point in the emotional cycle, it helps if you are in the early stage of the positive phase of the intellectual cycle as well when you stop smoking. The mind is ready to resist temptations and you are also physically able to cope.

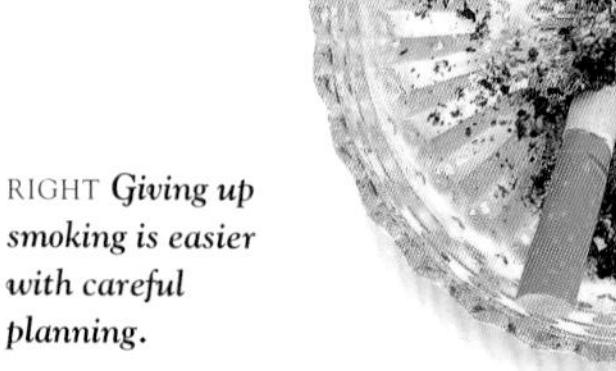

RIGHT ***Giving up smoking is easier with careful planning.***

WAIT FOR A CRITICAL DAY

On the first day after a physical critical day, preferably from positive to negative, enjoy one last cigarette.

Now! Stop smoking! If you can do it for one day, you can do it for two. After this, a week is nothing. By the tenth day, or nearly two whole weeks, your next physical critical day will come along. Don't worry about this because it does not last very long. After a month you will have stopped with the help of biorhythms. In the first few days your digestive system will tell you it wants more food and drink than usual. Any small weight gain is a part of giving up. Don't worry too much, just keep a healthy eye on your diet for the first few days or so.

RIGHT ***Make a determined start when you give up.***

To help increase your physical activity to offset this weight gain, park your car a little farther away from your destination and walk the rest of the way or, if you use public services and where it is possible to do so, leave the train or bus at a stop earlier than you normally would. Every little bit helps...

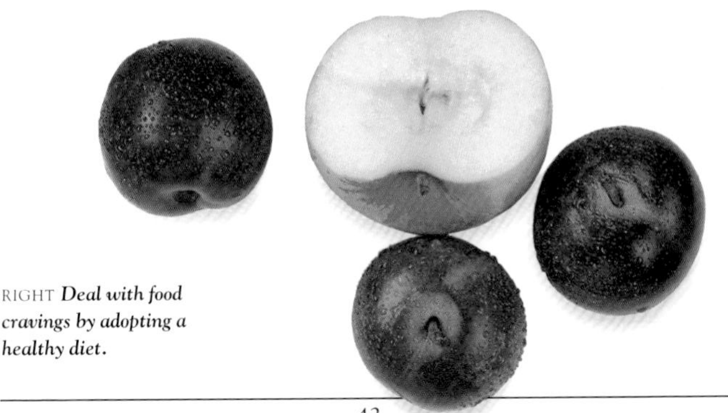

RIGHT ***Deal with food cravings by adopting a healthy diet.***

Win at sports with biorhythms

TO MAKE BIORHYTHMS WORK *for you in competitive sports you really do need to be aware of your three cycles at all times. Shooting and golf, archery and snooker require good visual skills. Football, swimming, running, and boxing all rely on the level of your physical stamina.*

All sporting activities demand a high triple biorhythmic state but this is obviously not always possible. A high physical cycle with either a high or low emotional or intellectual cycle needs to be carefully exploited for sports.

PLANNING IS ESSENTIAL

Chart your personal biorhythms well in advance of any sports program. Good planning is essential if you are to succeed. When all three cycles are in the plus phase you risk overdoing it. It's fine to have all the necessary energies at the right stage, but learn how to control and use them effectively. If all three rhythms are negative, then only your best efforts will count for anything. You won't set the world on fire but you can still give the opposition a good fight and win. When one or two rhythms are in their critical stages, your performance may be brilliant or erratic – or even both!

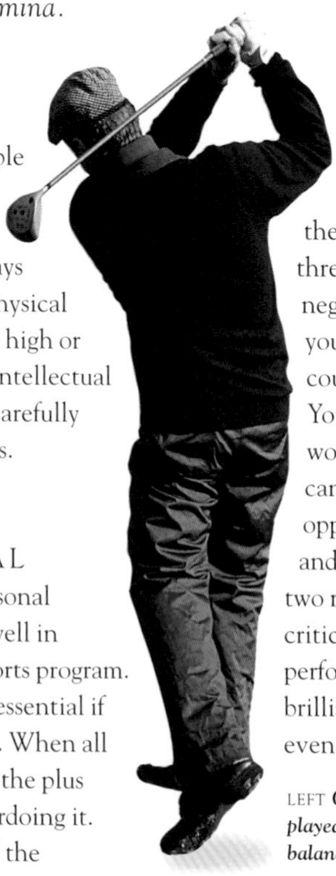

LEFT ***Golf is best played on days when balance is good.***

PLAYERS WHO WIN

Top tennis players who use biorhythms balance their low physical cycle with a positive stage intellectual cycle to cut out distractions. They rely only on themselves and their skills to survive. Adapt these guidelines to your preferred sport and you will start noticing the difference immediately!

In a competitive sport where you are part of a team, a positive emotional stage is helpful for sharing with others. If you are in a low physical stage you can only perform so far – avoid the temptation to overdo things.

These guidelines should help you to figure out what you need for any sporting activity not referred to here.

RIGHT ***Observe what your chart tells you.***

BELOW AND LEFT ***Some successful tennis players plan around their biorhythms.***

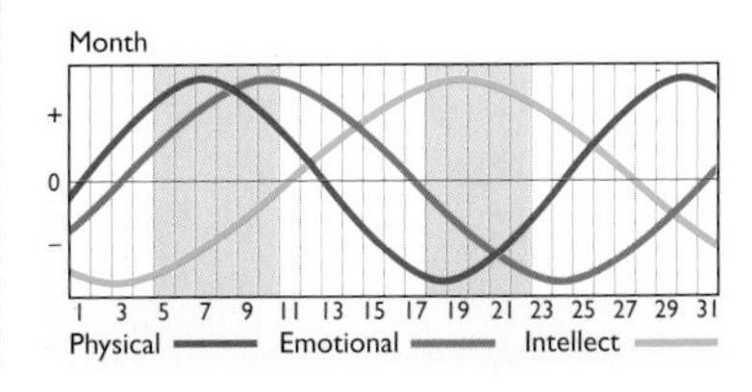

Business efficiency

BUSINESS EFFICIENCY *is a growth industry. Computer skills are now coupled with the specific technicalities of your job and are essential for survival. You need to stay on top or you become a loser. Timing, therefore, is vital.*

Travel, phone calls, e-mail, and meetings need management to be successful. There are days when you take all this in your stride. You make speedy decisions, have plenty of energy, and are efficient at managing time and people. On the other hand, we all have days when we feel under par, when our reactions aren't as quick as they need to be, or we aren't able to be as flexible as the ever-changing situation in the work place demands.

TAKE YOUR CHART TO WORK

To be a winner in today's world of work requires high personal performance all the time. Keep your biorhythm chart with your organizer and include, where possible, those of clients, managers, and your own staff.

RIGHT ***You can use the negative phases of the cycles to improve your work.***

Defer major decision-making on any critical day in case you make costly errors of judgment, and learn to use the negative stages of each cycle for handling routine matters at work.

In the positive phase you are able to grasp new ideas, have the strength to test them, and can take long business trips without overstressing your system.

LEFT *Draw up your clients' charts, if you can.*

CHECK YOUR CLIENTS' CHARTS

Plan meetings around the biorhythms of your business contacts, as well as your own. This is a growing tactic in business today as executives invest in their potential client's biorhythm calendar. They also check for compatibility. If their chart shows incompatibility with a client, they find a colleague whose biorhythm is more compatible and hand the casefile over. Deals are won on strategies like this. Think of the advantage you have over the opposition.

BELOW
Compatibility leads to deals.

Family relationships

WHEN A HYPERACTIVE CHILD *has a tantrum on a critical day the child is liable to hurt him- or herself. At the very least this will irritate the parents and, should one of you not be in a compatible mood, the result will be a bad time for all.*

ABOVE **On some days most families are naturally harmonious.**

RIGHT **You can predict your child's moods with his or her chart.**

Adults, especially parents, have a developed sense of awareness but children do not. Once you are aware of the effect biorhythms have on your children you can plan ahead to avoid difficult days.

AVOID DIFFICULT DAYS

Biorhythms provide a structured approach to inter-family relationships for, with the aid of your offspring's chart, you have advance notice of confrontational, hostile, or unproductive days and can learn to be more accommodating at those times.

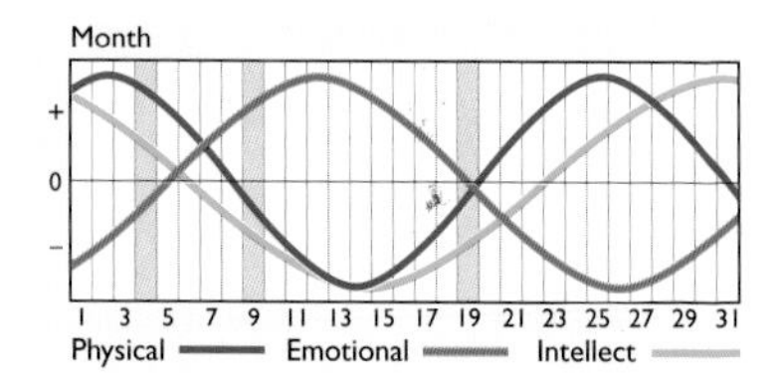

Parents do not have to change their overall attitude but simply adjust to the cycles and try to be more understanding of themselves and their children.

TALK OUT STRESS

It works in reverse, too: children begin to see their parents in a better light and understand that stressful situations can sometimes be talked through.

Prepare a compatibility chart for the family and mark off the danger days. Learning to use your biorhythms reduces domestic stress and strain and helps the whole family to live together in harmony.

LEFT ***Use biorhythms to check when your child is least accident-prone.***

COMPARING FAMILY BIORHYTHMS

You could use biorhythms to plan when to organize potentially stressful family events, such as moving house or going on vacation, by figuring out when the family's charts show that the positive cycles are most in tune with each other.

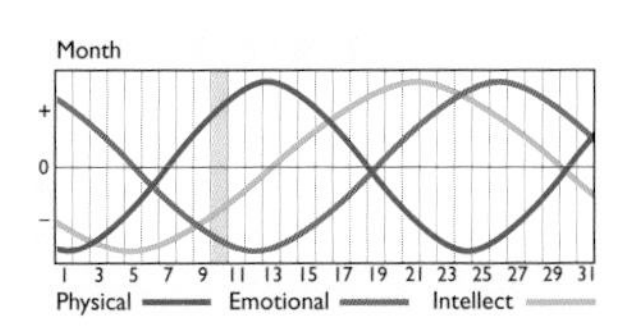

FATHER

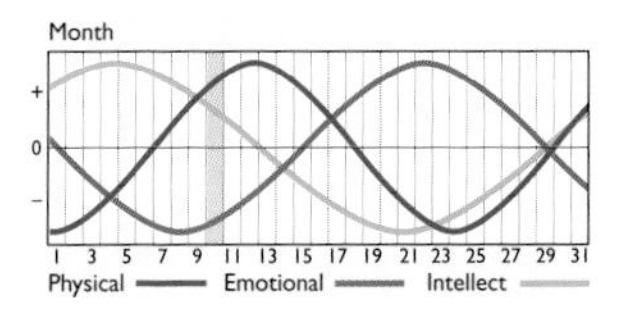

MOTHER

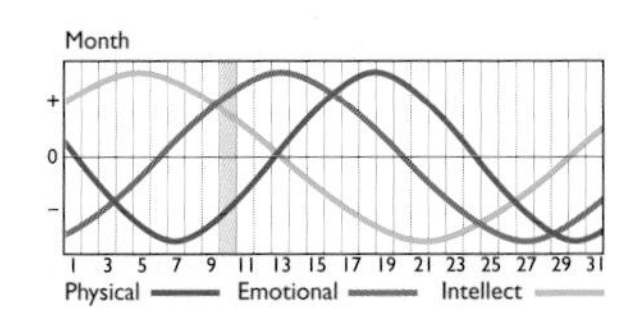

CHILD

Love relationships

THE START OF A NEW *relationship is a time of uncertainty for both partners as you gradually learn more about each other and try to anticipate each other's mood. Biorhythms can be especially helpful in this situation.*

LEFT ***Love needs more than romance to become permanent.***

Once you begin to understand and use biorhythms, you will find yourself quickly becoming sensitive not only to your own moods and abilities but also to those of others with whom you enjoy intimate relationships, especially lovers. When it comes to husbands, wives, and lovers, every change of mood and all their little personal gestures become high focus moments, and if you chart their biorhythms you'll soon recognize a pattern.

From a biorhythmic standpoint, a romantic or intimate relationship is superb for finding out about yourself and your partner and for defusing

BELOW ***Knowing your own chart will help you in the early days...***

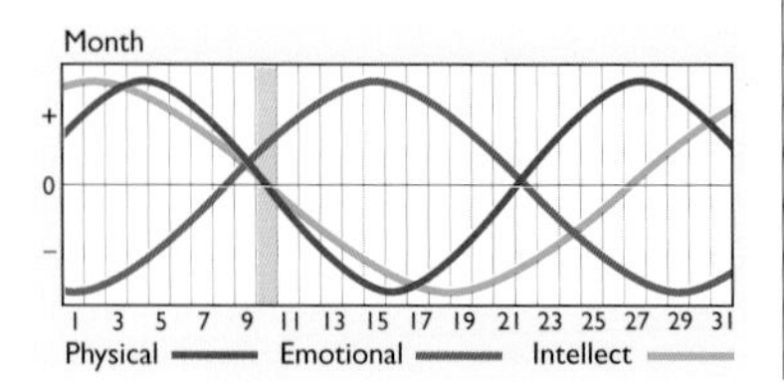

BELOW ***...while knowing your partner's chart can make you more sensitive.***

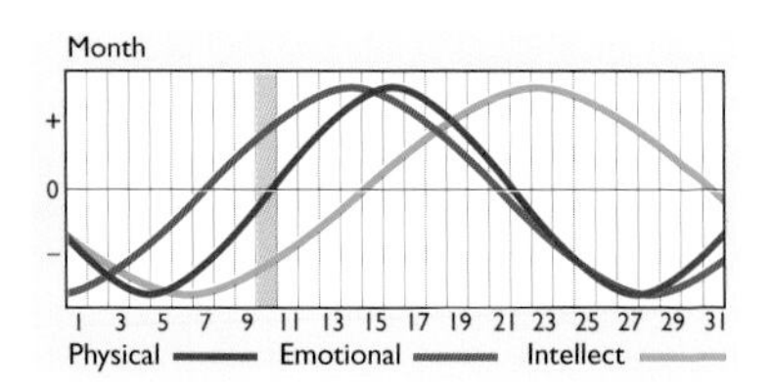

those moments of possible irritation that have the potential to sour a relationship.

As a rule, when you meet someone new it is often at a party or other kind of informal or formal get-together. You need to be emotionally stable to cope with these events. If they occur when your emotional rhythm is in the negative phase, you will have to work much harder to be sociable.

If your biorhythm is in the critical stage, you may be inclined to be a little oversensitive – the slightest thing could upset you. However, if your emotional cycle is in the positive phase there should be little reason for you not to shine at any social event.

Thus, when you meet someone new for the first time, neither of you is really being the real "inner" you. It is only later, when you are away from the artificial air of a party, that you really get to know one another properly. This is the time when it is worth taking the trouble to figure out your long-term biorhythm charts so that you can understand each other better and be more tolerant of each other's moods. This will give the relationship a better chance of success.

RIGHT ***It is after the first romantic encounter that you really get to know each other.***

Health, illness, and accidents

MEDICAL RESEARCH *asserts that biorhythms play a part in the development of illness. Incubation periods between exposure to a virus and the manifestation of it have been linked with the physical cycle critical days.*

HELP WITH SURGERY

Colds, flu, and other infections often tighten their grip on positive to negative physical critical days. On the other hand, people tend to recover more quickly from surgery in the positive stage of their physical rhythm. In surgery, bleeding is more profuse on physical critical days. Dental treatment is best carried out when all three cycles are in positive phases, for it is more painful when the physical rhythm is in the negative stage.

Heart attacks and strokes tend to occur when both the physical and sensitivity cycles are critical or in the low phase. It does not follow that this will happen, but it is more likely to happen when you are in poor health.

RIGHT ***Recourse to the doctor is more likely on critical days.***

WHEN ACCIDENTS MAY HAPPEN

There are some days and periods in the physical cycle when many people are vulnerable to accidents, particularly when the physical cycle is negative. Even athletic children used to a high level of physical activity may make a mistake or tire more easily at this time, making them much more likely to injure themselves. This is even more likely to happen when the intellectual cycle is positive – the combination may make them overconfident.

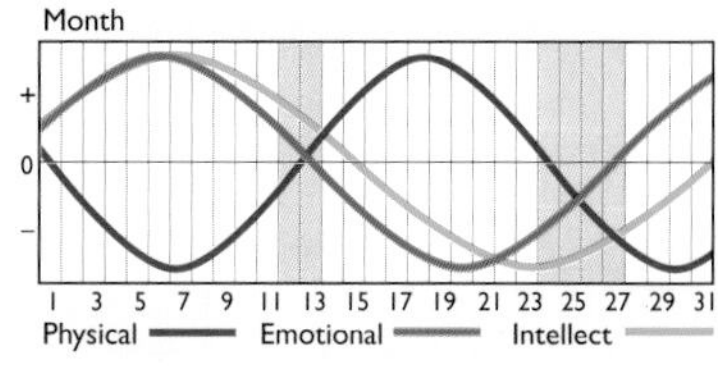

ABOVE ***This combination could lead to accidents.***

RIGHT ***When the physical cycle is negative, accidents are more likely to occur.***

ACCIDENTS CAN BE AVOIDED

Accidents frequently occur on single or multiple critical days when you are feeling over-confident. At this time it is best to be aware of your likely situation and avoid overstretching yourself. When rhythms are in the positive phase just prior to a critical day, you are likely to talk yourself into and out of all kinds of situations.

You are also likely to misread any warning signs if you are in a negative stage and fail to consider all the relevant facts before committing yourself to a decision, probably the wrong one. When this happens, it may not necessarily lead to an accident involving physical injury but it could be a business mistake that creates problems for you in other areas of your life.

Biorhythms in action

WE EXPERIENCE *six critical days a month, on the average, sometimes eight. If those six days occur every 30 days, then we are in a critical phase of our biorhythms for 20% of our lives, and for some people this can have significant consequences.*

Doctor Hans Schwing of the Swiss Federal Institute of Technology in Zurich published a paper in 1939 examining the state of biorhythms in the lives of people involved in accidents or accidental deaths. He calculated that 60% of accidents fall on critical days.

Doctor Rheinhold Bochow of the Humboldt University in Berlin came to similar conclusions in a paper he published in 1954.

Their figures suggest it is vitally important for us to be aware of our personal biorhythms. These four examples will show what I mean.

ABOVE ***Marilyn Monroe – victim of a biorhythmic state?***

MARILYN MONROE

BORN June 1, 1926

DIED August 5, 1962

OCCUPATION Movie Star

Born June 1, 1926, movie star Marilyn Monroe rose to fame in the 1950s, and her early death made her an icon. Although police reports claim she took a drug overdose on the night of August 5, 1962, the circumstances were suspicious.

But her biorhythmic state that night suggests otherwise. Her physical rhythm was in critical stage and she was about to have a double critical in her other two cycles. Any suggestion of suspicious death is not supported by the evidence shown here.

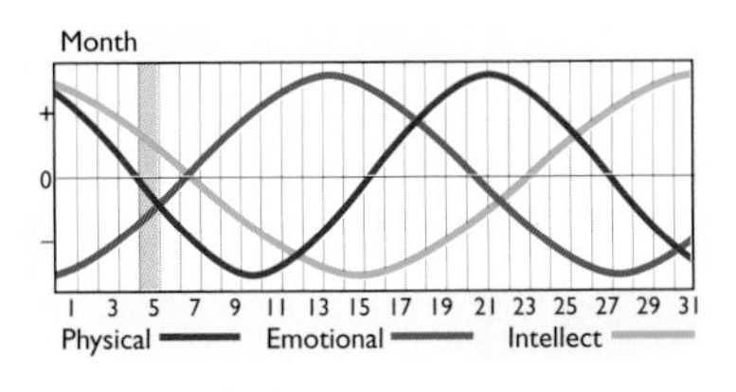

ABOVE ***Marilyn Monroe's biorhythmic chart at the time of her death.***

ELVIS PRESLEY

BORN January 8, 1935

DIED August 16, 1977

OCCUPATION Rock Star

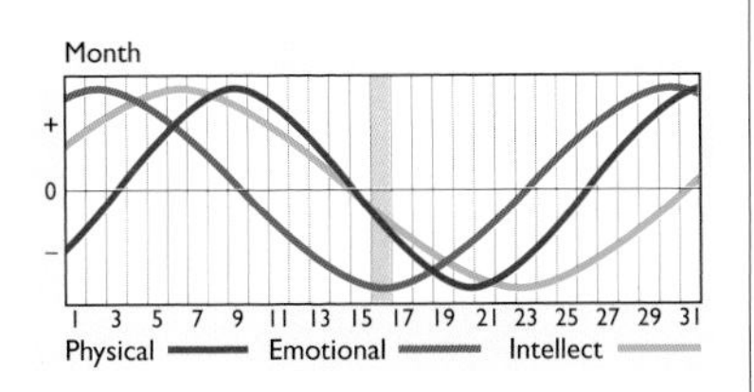

ABOVE ***Presley's chart shows negative states when he died.***

Elvis Presley collapsed at his home on August 16, 1977 and died later the same day. There is speculation that he may have committed suicide. Certainly, his cycles were all negative.

ABOVE ***Elvis Presley – died while in triple negative states.***

He had just had a double critical day from positive to negative stages in the physical and intellectual rhythms and his emotional cycle was in the negative mini-critical stage. We can conclude only that his biorhythms were at an extremely vulnerable stage.

ABRAHAM LINCOLN

BORN February 12, 1809

DIED April 14, 1865

OCCUPATION President of the USA

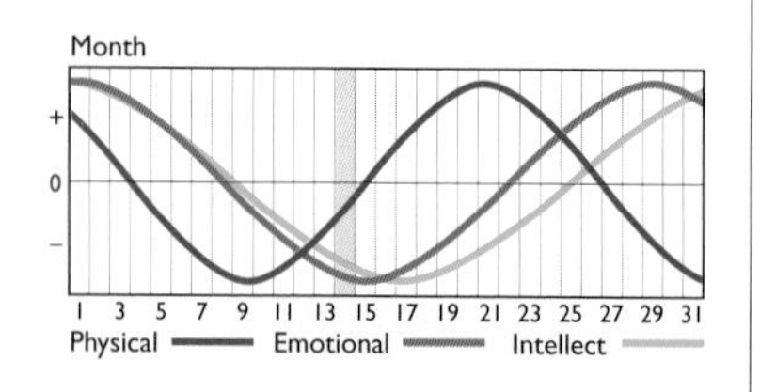

ABOVE *Lincoln may have ignored warnings the day he was shot because his biorhythmic state inclined him to recklessness.*

On April 14, 1865, while watching a play at Ford's Theater, President Lincoln was shot in the head at point-blank range. Although friends had warned him several times to take safety precautions, he chose to ignore them.

It was less than a month after he had given his second inaugural address in which he had discussed the profound moral significance of the Civil War. The speech aroused a lot of animosity and, on that terrible Good Friday, John Wilkes Booth exacted his revenge. This was all the more tragic because Lincoln, possibly because of his own personal fairness, morality and Christian values, would not accept that he had enemies who would go that far. He dismissed as unnecessary and unwarranted, the warnings that he ought to have a bodyguard.

Five days earlier he had experienced a double positive to negative critical day emotionally and intellectually and was in a critical stage physically. This might have contributed to the disregard he paid to the warnings and led to his violent end.

INDIRA GANDHI

BORN November 19, 1917

DIED December 31, 1984

OCCUPATION Prime Minister of India

ABOVE *Indira Gandhi, India's Prime Minister, assassinated in 1984.*

Born November 19, 1917, Indira Gandhi became the first woman Prime Minister of India in 1966 but was assassinated on December 31, 1984 by members of her Sikh bodyguard. She was a headstrong personality, known to ignore advice from security staff that might have saved her life on this occasion.

But at the time her emotional rhythm was in a critical stage and the other two cycles were in a negative phase. With her biorhythms in this state, she may have felt overconfident and failed to respond to the warnings.

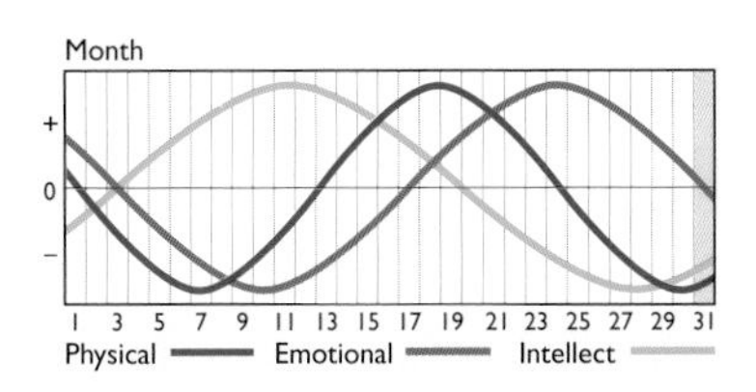

ABOVE *Indira Gandhi's normal intransigence was exacerbated by her biorhythms on the day of her death.*

SUMMARY

The catastrophic deaths of these four people help to validate biorhythm theory and practice. Examinations undertaken on the biorhythmic states of historic figures reveal that their actions or decisions might have been influenced by their biorhythms.

Once we are aware of our own personal cycles, we can begin to experience a more satisfactory way of life in every sense. In effect, we learn to live with and adapt to our biorhythms.

Creating charts

EACH BLANK CHART *represents a month of 31 days, so adjust for the shorter months. Please feel free to make photocopies of this page of blank charts to help you work as accurately as possible.*

Always make your basic calculations for the first day of a month because it makes it easier to progress the rhythm right through that month.

To create your biorhythm chart you will need to use three pens – red, blue, and green. Use red ink to record the physical rhythm; blue ink to show the emotional or sensitivity cycle; and green ink to draw the intellectual rhythm.

A small protractor, a calculator, and some scrap paper for notes and calculations are all the other basic tools needed.

Once you have established where each cycle is on the first day of the month, indicate where the critical days occur with a small dot. Link the dots using the protractor with the right ink and in the correct direction, positive above the line, negative below the line.

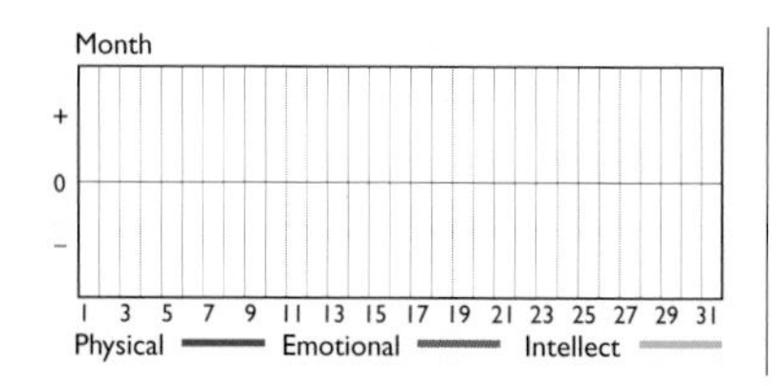

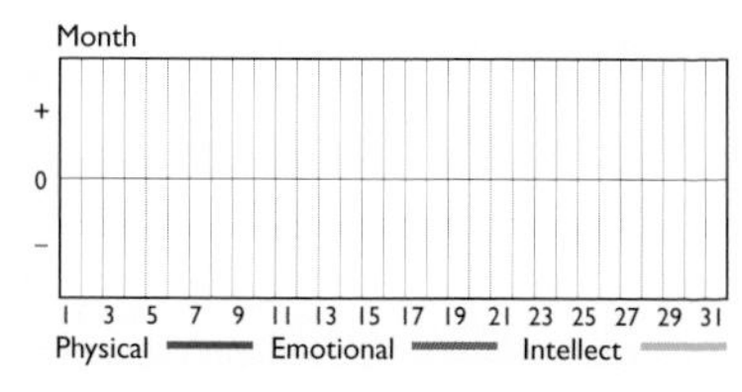

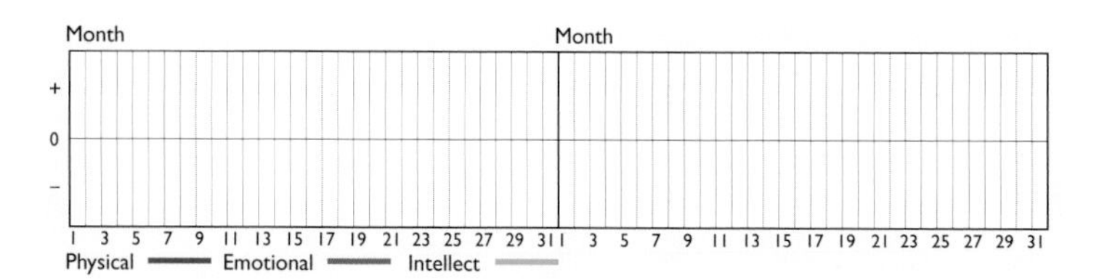

Further reading

WINSTON CHURCHILL'S AFTERNOON NAP
by *Jeremy Campbell*
(Aurum Press, 1988)

THE BIORHYTHM BOOK
by *Jacyntha Crawley*
(Virgin Books, 1996)

CYCLES
by *Edward Dewey*
(Hawthorn Books, 1971)

BIORHYTHMS, A PERSONAL SCIENCE
by *Bernard Gittelson*
(Arco Publishing, 1975)

BIORHYTHMS AND YOUR BEHAVIOUR
by *Vincent Mallardi*
(Running Press, 1975)

BIORHYTHM & YOU – THE FACTS
by *Don Rebsch*
(UBC, 1977)

IS THIS YOUR DAY?
by *George Thommen*
(Crown Publishers, 1974)

BIORHYTHMS
by *Peter West*
(Thorsons, 1980)

Other Books in this Series

ALEXANDER TECHNIQUE

ALTERNATIVE MEDICINE

AROMATHERAPY

BACH FLOWER REMEDIES

CHINESE HERBAL MEDICINE

HERBAL MEDICINE

HOMEOPATHY

MASSAGE

NATURAL BEAUTY

NATURAL HOME REMEDIES

REFLEXOLOGY

SHIATSU

VITAMINS AND MINERALS

YOGA